POWER SHIFT

The hope and drama of monumental wind power

POWER SHIFT

DONALD A. PETTIT

Published in Canada by Peace photoGraphics Inc.
1204 - 103rd Avenue, Dawson Creek, British Columbia, Canada V1G 2G9
phone: 250-782-6068 toll free: 1-888-373-8488
email: info@peacephotoGraphics.com
www.peacephotoGraphics.com

Produced by Peace photoGraphics Inc. using both solar and wind power
Principal photography by Donald A. Pettit
Additional photography by Aeolis Wind Power Corporation, Hank Bridgeman,
Terry Bryant, Kit Fast, Andrew Kohl, Hugh McNair
Written and designed by Donald A. Pettit
Edited by Barbara Swail
Illustrations by Barbara Swail, Tracy Wandling
Printed and bound in Canada by Friesens Corporation
Printed on FSC certified paper with vegetable-based inks

Mixed Sources
Cert no. SW-COC-001271
© 1996 FSC

Library and Archives Canada Cataloguing in Publication

Pettit, Donald A., 1948-
    Power shift : the hope and drama of monumental wind power / Donald A.
Pettit.

ISBN 978-0-9736678-8-2 (pbk.).--ISBN 978-0-9736678-9-9 (bound)

    1. Bear Mountain Wind Park--Pictorial works. 2. Wind power
plants--British Columbia--Dawson Creek--Pictorial works. 3. Wind
turbines--Design and construction--Pictorial works. I. Title.

TK1541.P48 2010   333.9'20971187   C2010-903900-9

Acknowledgements

I gratefully acknowledge the support of Peace Energy Cooperative, Aeolis Wind Power Corporation, ENERCON Canada Inc., Eagle West Wind Energy and Hatch. AltaGas Income Trust has generously released the construction photos of Bear Mountain Wind Park for use in this book and provided the expertise to ensure accuracy. In particular, I wish to thank Marc Boucher, whose technical knowledge seems to be without limit, and Jim Bracken, who has been so open to creative ideas and unreserved in helping make them happen. To everyone at AltaGas, I thank you. You have been a pleasure to work with.

On the work site I am indebted to Dan Irving and Jay Walker who were immensely tolerant of my impetuous enthusiasm and who patiently answered thousands of questions during their busy work days.

Peace Energy - A Renewable Energy Cooperative's board and executive director, Valerie Gilson, have been long term supporters of my work and are enthusiastically helping with the promotion and distribution of this book. The members of Peace Energy Cooperative deserve high praise and special mention for their role in conceiving of Bear Mountain Wind Park and shepherding it through to completion while adding the critical component of local ownership.

A long list of people provided critique, research, expertise and proof reading: Marc Boucher, Don Hallam, Adrianne Lovric (AltaGas); Colleen Brown, Juergen Puetter, TJ Schur (Aeolis); Marc-Antoine Renaud (ENERCON Canada Inc.); Murray Westerberg (Eagle West Wind Energy); Dan Irving, Dan Perrin (Hatch); Valerie Gilson (Peace Energy Cooperative); Kit Fast, Cees van de Mond and Tracy Wandling (Peace photoGraphics Inc.).

Most of all I am indebted to Barbara Swail, for her unfailing support for yet another of my book obsessions. She is a skilled editor and illustrator, and an honest critic (rare!). Thank you, again.

Previous spread: *the rugged ENERCON turbines of Bear Mountain Wind Park were working hard, turning boldly in the strong north wind of this March 2010 blizzard.*

A chorus line of wind turbines
posed for Chinook's embrace

once given, she snowdrifts
wildly into the north

her now, now, now
and here, still
scented with ocean

Marilyn Belak

Contents

For eons uncountable, a constant wind has flowed out of the Rocky Mountains and across the Peace River region at the northwest edge of Canada's grand prairies. Pine Pass funnels the steady blow toward Bear Mountain, a ridge that juts northward across the air stream, 15 kilometres southwest of Dawson Creek, British Columbia, Canada.

Today, 34 three-megawatt ENERCON wind turbines form a sinuous line along the crest of Bear Mountain ridge. Beneath it spreads the rolling fertile farm and ranch land stretching east to Alberta's prairies and west to a line of distant snow-capped mountains. This is Bear Mountain Wind Park, commissioned in October 2009, British Columbia's first utility-scale wind development. At 102 megawatts peak output, it produces enough pollution-free electricity to run 35,000 North American homes, or every household, business, factory, street light and cell phone in British Columbia's South Peace region, from Dawson Creek to the distant

mountains.

More than quantity, the region's wind is characterized by quality: mostly from one direction, low in turbulence and with high seasonal consistency. With a wind resource estimated to exceed 10,000 megawatts, the Peace region is rich in a new and vital resource of global significance.

The story of Bear Mountain Wind Park is the story of how state-of-the-art wind turbines are built, how they work and why they are critical to our future. It is the story of a visionary grass-roots group, a community-minded development firm and an innovative energy corporation and how they worked together in new and ground-breaking ways to bring the first wind power to British Columbia.

This book tells the story, in pictures and words, of how and why this came to be.

Foreword

"We've arranged a civilization in which most crucial elements profoundly depend on science and technology. We have also arranged things so that almost no one understands science and technology. This is a prescription for disaster." **Carl Sagan**

A dream assignment landed on my desk: document the construction of a wind park going up near my city. It was Bear Mountain Wind Park, the first wind utility to be built in British Columbia.

It began in 2004 when Peace Energy Cooperative (PEC), a community-based organization I helped found, secured the license to assess the possibility of utility scale wind on Bear Mountain. I helped with the search for a developer who would respect the park-like character long cherished by locals and embrace our fledgling cooperative as a partner. So when the day finally arrived to break ground, I was thrilled to be there.

After years of dreaming and endless meetings, it was amazing to watch the project get underway with swift efficiency. Proving the site,

assessing environmental concerns and finding partners and financing were now behind us. Finally the turbines were shipping from the German company ENERCON and the towers from Hitachi in Saskatoon, Saskatchewan.

PEC members watched via Internet as the first ship load of blades and components traversed the Atlantic, navigated the St. Lawrence Seaway and Great Lakes, then were loaded on special trains and trucks in Thunder Bay for the long journey across Canada.

On a raw and blustery Monday morning in May, everyone was tucked indoors. But when the blades began to arrive in Dawson Creek, word went out from household to household and vehicles pulled up as the train pulled in, with PEC members spilling out to

hoot and high-five the arrival.

Experts from around the world were also there to meet the shipment. They watched our enthusiasm with amusement. "Why are you taking so many pictures?" one asked with a thick German accent. "They are only wind turbines." To them, wind energy was nothing new, of course. ENERCON has manufactured turbines for more than 25 years and installed thousands worldwide. But we had hundreds of hours of hard work invested in bringing wind power to our region. For us, this was a red-letter day.

Over the two years that I photographed the construction of Bear Mountain Wind Park, I met many people who worked there, from surveyors and biologists to cement workers, engineers and CEOs. Their pride in the project was apparent.

They knew wind power to be a key component of the new world we must create as quickly as we can, and, as part of the team building British Columbia's first wind utility, they were making history.

Important work but also routine. This was monumental wind power on a mass-production scale: build the roads, pour the foundations, erect the towers, assemble the components, fire 'em up, all with clockwork precision, then on to the next.

What was their motivation? Simply put, it was just their job. And this is the way it has to be if the POWER SHIFT, the rapid switch to clean energy, is going to happen any time soon. Corporate and government leaders have an immense role to play in making every job a "green" job, so that solving our global problems and

building a better world just means getting up and going to work in the morning. Up on Bear Mountain, I saw it in action. I know it can work.

As a photographer, I welcome these gentle giants to our Peace Country landscape, one that I have explored, photographed and grown to love over three decades. To me they are beautiful, utterly functional yet supremely elegant.

In my view, wind power is one of the greatest opportunities we have today, and those who earn their livelihoods in this industry are unsung heroes. This book is a record of what they accomplished near one small city named Dawson Creek. It is a tribute to them all.

It is my hope this book informs and inspires: that you learn something about wind energy you didn't know before, and that you feel the thrill of the POWER SHIFT now sweeping the planet. We have the technology and the skills to build a better world. All we need now is the will and the wisdom to ask for a little help from the wind.

Don Pettit

Above: *Dan Irving was construction site superintendent for Bear Mountain Wind Park and my guide for the two-year photo shoot that made this book possible. Dan had four wind projects under his belt, all in southern Alberta ranch country. I thought of Dan as "a wind power cowboy" – quiet-spoken, down to earth and very strict on worksite safety. He always seemed to be in a good mood, but not the kind of guy you would want to cross. In short, he was the ideal superintendent and as a guide for a wide-eyed photographer, unbeatable.*

Facing page left: *the turbines of Bear Mountain Wind Park waiting for the calm weather that will allow their blades to be lifted into place.*
Facing page right: *author and photographer Don Pettit on assignment, documenting the construction of Bear Mountain Wind Park.*

The towers of Bear Mountain Wind Park catch the warm gleam of the setting sun at 3:30 in the afternoon on a cold December day in 2009. Nearby Dawson Creek lies hidden in the rolling hills between the photographer and the distant ridge.

Wind power:
A Short History

Above: the classic Dutch windmill was used for more than 800 years to pump water off fertile soil beneath lakes and seas in an effort to increase the area of the Netherlands. They harnessed the mechanical power of the wind for many industrial purposes, including grain milling and saw milling.

Above left: these Cretan antiques are surviving examples of windmills that covered Europe and the Mediterranean for hundreds of years, grinding grain and drawing water. Above right: fifty years after its invention in 1854 by Daniel Halladay there were an estimated six million wind powered water pumpers in use on the prairies.

The Persians are well known for conquering the Mediterranean in wind-powered sailing ships. Few realize they were pumping water, grinding grain and cooling buildings with wind machines – 3,000 years ago! There is evidence of wind-powered water pumps and grinders in China at that time, as well.

The earliest wind machines were vertical axis, the simplest to build and operate. By the 11th century, horizontal axis windmills, the precursors to today's giant turbines, were well distributed throughout the Mediterranean and Europe. Working examples still exist today, including more than 1,200 iconic Dutch mills.

The first electricity-generating windmill was built by James Blyth of Glasgow, Scotland in 1887. The next year, a larger machine designed by Charles F. Brush was operating in Cleveland, Ohio. Thousands of Canadian and American prairie farms were electrified with wind as early as the 1930s, long before there was a rural electrical grid. The two- and four-bladed machines can still be found, refurbished and fully functioning.

By 1900, Denmark had about 2,500 small windmills with a combined capacity of 30 megawatts (MW), largely thanks to scientist and inventor Poul la Cour. Today, about 30 per cent of Danish power comes from wind and the country is the world leader in production, design and manufacture of wind generators.

The world's first megawatt wind turbine, a massive 1.25 MW wood-towered generator was installed on Grandpa's Knob in Vermont in 1941, but the Danish company Vestas deserves the credit for taking wind to utility scale. A 30-kilowatt (kW) Vestas, the first large-capacity turbine intended for mass-production, was

Above left: *Jim Toth works on a Wincharger refurbished by Terry Bryant at his farm in the Amish country of western New York State.* **Above right:** *the 218-tonne turbine built in 1941 and installed at the summit of Grandpa's Knob in Vermont was the first to break the megawatt barrier, proving that large-scale wind power was possible.*

Above: *construction of the first wind farm in the U.S. was started in 1981 in Altamont Pass, California. By 1986, it had 576 MW capacity, almost half of all the wind-produced energy in the world, with generators as small as 50 kW. Today, many of the original 6,900 small generators have been replaced with much larger, more efficient units.*
Right: *The ENERCON E-126, the world's largest wind turbine, is rated at 7.5 MW. Fifteen are in operation in Germany and Belgium, with more on the way. They feature the same gearless ring generator pioneered by ENERCON and detailed in this book. These behemoths are 135 metres high at the hub with a rotor diameter of 126 metres.*

installed in 1979. By 2004, Vestas dominated the industry globally, claiming 34 per cent of installed capacity worldwide.

Many companies and designs have emerged over the past 30 years as wind has become both financially competitive and the best solution to low carbon electricity. The ENERCON E-82s featured in this book, rated at three megawatts, are at the cutting edge of power and efficiency. With

34 E-82s, Bear Mountain Wind Park is the largest single installation of this turbine. ENERCON also makes the world's largest operational wind turbine, the 7.5-MW E-126 (photo at right), although the race to maximize capacity and efficiency continues.

Wind was once the world's primary energy source. With global investment in wind energy projected to exceed $1 trillion by 2020, we are quickly coming full circle.

Wind turbines:
How they Work

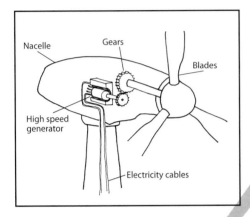

Most wind turbines use gears to increase rpm to drive a generator housed in the nacelle. This design has been used very successfully for decades by many turbine manufacturers.

FAST FACTS about Bear Mountain Wind Park

- **Turbine:** ENERCON E-82 E3, manufactured in Germany
- **Rated power:** 3 megawatts each x 34 = 102 MW total
- **Turbine design:** gearless, variable speed and pitch control
- **Blade material:** fibreglass-reinforced epoxy resin with integrated aluminum lightning protection
- **Rotor:** 82 m diameter, upwind clockwise rotation
- **Speed of blade tip:** 25-80 metres/sec. (6 – 19.5 rpm)
- **Cut-in, cut-out wind speeds:** 2.5 m/s, 22 - 28 m/s
- **Tower:** 4-section, steel, manufactured in Saskatchewan
- **Tower weight:** 214 tonnes
- **Hub height:** 78 m
- **Remote monitoring:** 24 hour ENERCON SCADA

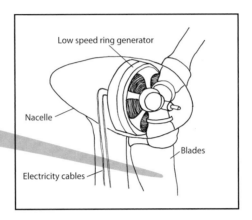

The ENERCON design is unique because it does not use gears to increase rpm. Instead, a large-diameter ring generator attached to the front of the nacelle is driven directly by the rotating hub and blade assembly.

LONG BLADES

Longer blades increase power output. As the size of the "swept area" of the blades increases, more energy can be extracted from more wind. The effect is dramatic: doubling blade length increases power by four times.

TALL TOWER

Wind moves faster higher up than at ground level. Doubling the height of a wind turbine increases the energy it can produce by more than a third.

Total height → to blade tip: 119 m

Wind Direction

Hub ← Height 78 m

Section 4
Length: 24 m
Weight: 35 t

Section 3
Length: 21.8 m
Weight: 48 t

Section 2
Length: 18.2 m
Weight: 55 t

Total tower weight: 214 t

Section 1
Length: 12.35 m
Weight: 61 t

Foundation Section
Length: 2 m
Weight: 15 t

Inside the E-82 E3

Each turbine site has unique air flow patterns over the surrounding terrain. Each wind turbine must therefore work autonomously, constantly adjusting to face the wind and changing blade angle to maximize energy conversion.

Wind speed and direction data is collected from sensors on top of the nacelle and fed to the on-board computer which controls motors that rotate the nacelle into or out of the wind. Other motors rotate the blades individually to maximize power output while minimizing load on the system, or to turn the blades out of the wind to stop the turbine completely or put it into a slow-rotation "idle" mode. An ice detection system will automatically shut down the turbine if ice or hoar frost forms on the blades.

This fine tuning is continuous and automatic. On-board back-up batteries are held in readiness to power all systems should there be a grid failure.

Although each turbine operates autonomously, all are monitored continuously from several remote locations by supervisors using the ENERCON Supervisory Control and

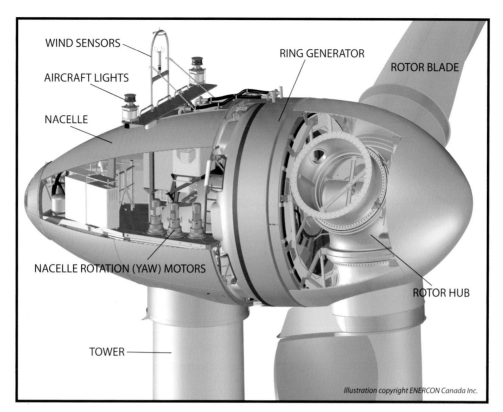

Illustration copyright ENERCON Canada Inc.

Above: *a large ring generator sets the ENERCON design apart from all others. It is driven directly by the hub and rotor assembly with no intervening gear box.*

Data Acquisition (SCADA) system. All turbines can be operated manually or shut down entirely from each location. Power output is recorded, problems identified and repair crews dispatched using the SCADA system.

Most control, monitoring and power conditioning equipment is housed in the base of the tower (the E-module) for easy access. Should repair or maintenance work be needed in the nacelle, a ladder or

winch-driven person lift is used to ascend and descend the tower. A hatch in the lower rear of the nacelle allows equipment and tools to be lifted via winch from the ground and also provides emergency egress.

The ENERCON design, with its low-speed direct drive ring generator, is considered by many to be the quietest, most efficient and most reliable wind generator in the world.

Above: *a service crew winches equipment into the nacelle through the rear access hatch. For safety reasons, work can only be performed in the nacelle when rotation is stopped by feathering the blades and breaking to a full rotor stop, as seen here.*

How to build a wind park

1. Find wind

Wind prospecting has become a world-wide industry. Using sophisticated topographic and atmospheric computer models, prospectors narrow their search, then travel to likely sites to stake claims which can be worth millions if the site "proves out." Road access to the potential site, availability of local services and equipment, and proximity to the grid are all critical factors.

It is not just quantity, but quality of wind that is measured. Steady, reliable wind from one direction is better than a much stronger wind that is constantly changing direction and intensity. Extreme weather of all kinds can temporarily shut down operations.

2. Monitor the site

Towers placed at key locations collect data about the actual wind resource. Meteorological data must be collected for at least one year, preferably longer, and certified as accurate. The expense and difficulty of erecting towers in remote locales may eventually be reduced as new wind monitoring technologies that use sound waves, lasers or radar become more reliable and accurate.

3. Build a business model

Wind projects, large or small, must be profitable. That means finding partners with the money and experience to fund and build them on time and on budget. It also means negotiating energy purchase agreements with utilities, negotiating leases with land owners, and much more. All of these key business pieces must fall into place, or the project will not move forward no matter how good the wind resource.

4. Perform an EA

In British Columbia, the Environmental Assessment (EA) process for wind

Left: a giant hub assembly weighing 24.5 tonnes is readied for lifting to the top of the 78 metre tower. Three blades will be attached to the hub at the blade adapters (wrapped in red). The assembly rotates and drives the generator.
Right: the second steel tower section, weighing 55 tonnes, is lifted into place atop the first tower section at the far left.

projects is exhaustive. Detailed wildlife studies, often running for years before project approval, during construction and after project completion, are required, as well as air and water impact assessments, archeological surveys, and much more. Extensive public information and consultation are also required. (for more about the Bear Mountain Wind Park EA see pages 28-29).

5. Build

With tens of thousands of wind turbines installed world wide, building a wind park has become relatively routine. For the Bear Mountain Wind project, access roads, power lines and turbine foundations took roughly one year to build. Erecting and commissioning the wind turbines took another six months. Overall, some 400 people were employed, about 300 on-site.

6. Operate

The operational lifetime of a modern wind park is usually 25 years. For a project like Bear Mountain Wind Park, one operations supervisor and six service technicians (four electrical, two mechanical) perform routine maintenance and trouble shooting. The maintenance downtime on a

modern wind turbine is less than two per cent per year, matched for reliability only by photovoltaics, and is much better than, for instance, a coal-fired plant at 12.5 per cent.

The 'capacity factor' is the amount of electricity actually produced as a proportion of the generator's output if it operated at maximum power all the time. Because the wind is a variable power source, the capacity factor for wind turbines is between 25 and 40 per cent, with the Peace region expected to be at the high end of that scale. For comparison, hydro dams run at 45 to 50 per cent and coal plants at 70 per cent capacity factor.

7. Recommission

After 25 years, significant refurbishing of blades and generators will be required if the wind park is to continue operation. Some steel towers may need replacement, depending on the loads they have been subjected to during operation. However, much of the infrastructure, such as the cement foundations, should be serviceable much longer than 25 years. It is generally agreed that with the installation of refurbished equipment and new technology, the wind park's lifetime could be extended almost indefinitely.

Above: *although the design of turbine foundations varies with soil type and many other factors, they must be immensely strong and stable. Here, 300 tonnes of concrete is poured into the network of heavy reinforcing steel to create the wide, flat foundation most suitable for the Bear Mountain site.*

Right: an aerial view of Bear Mountain Wind Park nearing completion in late summer of 2009. Almost daily a turbine is finished and brought on-line, slowed only when high winds make crane work dangerous. As each turbine is tested and spun-up, it begins to feed power into the grid.

PLANNING and PUBLIC AWARENESS

Peace Energy Cooperative

In 2001, two Dawson Creek entrepreneurs (the author and Paul Kurjata) began discussing the formation of a cooperative that would promote the development of renewable energy in the British Columbia and Alberta Peace region. Peace Energy - A Renewable Energy Cooperative (PEC) was incorporated in October 2003, a for-profit cooperative that was the first of its kind in western Canada, and one of only a handful in North America.

PEC's goals were to make renewable energy both practical and profitable and to attract a membership large enough to raise the funds to venture a project on its own or attract a partner with the resources. Almost immediately, a remarkable opportunity presented itself.

Wind prospectors had discovered the Peace region and were beginning to quietly stake claim to its rich wind resource. British Columbia's publicly owned utility, BC Hydro, had been collecting wind data from Bear Mountain, near Dawson Creek, for about a year, but was shifting its mandate to purchase new power from independent power producers (IPPs) rather than building its own infrastructure. In 2004, PEC acquired BC Hydro's wind monitoring equipment on Bear Mountain and looked at the data: the numbers were encouraging. Bear Mountain could be a world-class wind location.

In June 2004, PEC acquired the crown land investigative use permit (IUP) to evaluate Bear Mountain's commercial wind potential. The search for a developer led to Aeolis Wind Power Corporation of Sidney, B.C. Aside from their expertise, Aeolis was chosen for their commitment to the region and to the cooperative model. PEC and Aeolis signed a joint working agreement in October 2005.

In December 2005, Aeolis signed agreements with AltaGas Income Trust of Calgary, Alberta that lead to the formation of Bear Mountain Wind Limited Partnership (BMWLP). The participation of AltaGas ensured there was money and experience to complete the project. In July 2006, BMWLP was awarded a 25-year energy purchase agreement with BC Hydro, and the environmental assessment was approved in August 2007, with ground breaking in December.

From the beginning it was clear that PEC brought value to the project

Above top: Peace Energy Cooperative members had to get up to speed on the wind industry very quickly. PEC's first manager, David Kidd, traveled with other members to Pincher Creek, Alberta in 2003 to have a look at existing wind facilities first hand. *Bottom:* PEC founding members Cees van de Mond and Barbara Swail helped spread the word about the coming Peace Country "wind rush".

in the form of public support and member investment. Membership grew steadily to almost 400 by project commissioning in October 2009. With a retained financial interest in BMWLP, the cooperative, mature, well-supported and successful, was ready to move forward with other innovative renewable energy projects in the Peace region.

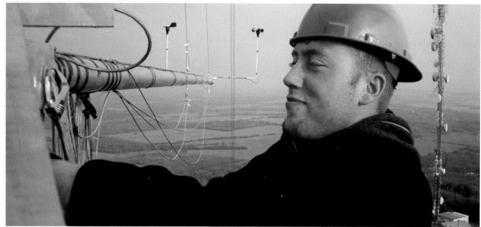

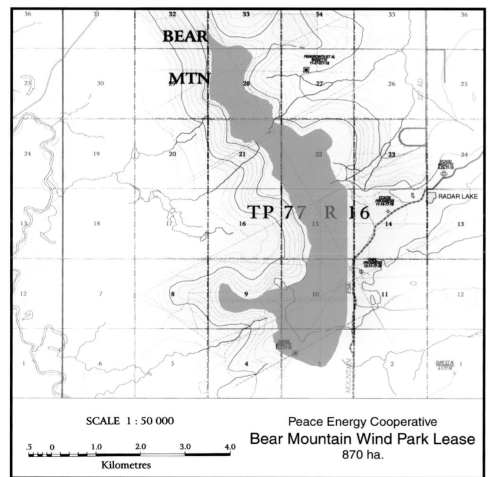

SCALE 1 : 50 000

Peace Energy Cooperative
Bear Mountain Wind Park Lease
870 ha.

Kilometres

Above and left: *early wind data collected from equipment mounted on existing communications towers, first by BC Hydro, then PEC and then by Aeolis Wind Power Corporation, indicated that Bear Mountain could be a world-class wind power location.*

Left top: *the first crude simulation (by the author, for PEC) was used to show the approximate scale and look of a wind park on Bear Mountain.*
Bottom: *once PEC acquired the crown land investigative use permit (IUP) in 2004, the cooperative prepared this map to show the location and extent of the intended development.*

Previous page 22: *a concept drawing by Barbara Swail illustrates the vision of the completed wind facility as a "park" that would enhance historical grazing and recreational uses with added educational components.*
Previous page 23: *this photo simulation by Tracy Wandling and the author shows mature regrowth and naturalization of the site in an accurate scale representation of the park.*

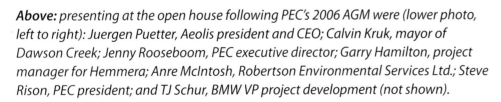

Above: presenting at the open house following PEC's 2006 AGM were (lower photo, left to right): Juergen Puetter, Aeolis president and CEO; Calvin Kruk, mayor of Dawson Creek; Jenny Rooseboom, PEC executive director; Garry Hamilton, project manager for Hemmera; Anre McIntosh, Robertson Environmental Services Ltd.; Steve Rison, PEC president; and TJ Schur, BMW VP project development (not shown).

Above top: Jenny Rooseboom, PEC executive director, answers questions at a Dawson Creek public information session in 2005.
Bottom: Greg Dueck and other volunteers manned a display at a local mall the same year. PEC used every opportunity to inform the community and attract new members. They found the public to be keenly interested in renewable energy.

Above top: Marc Boucher, AltaGas project manager, points out features on a project map. Community interest was high and the companies involved provided experts and information at a series of open houses.

Bottom: Jim Bracken, AltaGas senior vice president, major projects, is interviewed by the media in Dawson Creek.

Above top: by November 2006, more than 50 meetings and information sessions had been held with stakeholder groups, elected representatives and the public, including many open houses like this.

Bottom: PEC members Greg and Joanne Dueck and family with Valerie Gilson (right), PEC executive director, at the 2006 Dawson Creek Kiwanis Trade Fair.

The Environmental Assessment

W ind parks, like other major developments, are subject to an Environmental Assessment (EA) under the British Columbia Environmental Assessment Act (BCEAA) and the Canadian Environmental Assessment Act (CEAA). The intent of the EA process is to identify and address any foreseeable adverse effects throughout the life cycle of a project – including construction, operation and decommissioning – and to determine ways to avoid, eliminate, minimize or mitigate those effects. The EA also provides meaningful opportunities for consultation with all interested parties.

Ecological Community Planning

Ecological Community Planning (ECP) sets the highest standard in project development and community consultation. Developed by TJ Schur, VP project development for Bear Mountain Wind, ECP is a planning model designed to encourage community investment and participation. As preliminary work continues and grows on the technical side, equal importance is placed on recognizing and establishing opportunities and benefits to the local community. This includes local capacity building, identifying and establishing local opportunities, and forming educational, grass-roots and regulatory connections with local support early in the process. While much of this work is also part of the EA process, with ECP these partnerships and relationships are sought out early on and continue not only through the EA but also through the life of the project.

Peace Energy Cooperative and Aeolis began working with local citizens, regulators, First Nations and special interest groups in June 2004, more than a year in advance of the EA process which has its own public and First Nations consultation requirements. Prior to initiating the EA in November 2005, Aeolis and Peace Energy Cooperative had held over 14 meetings with local stakeholder groups and organizations and countless individuals in the region. By November 2006, more than 50 meetings had been held. Public access to information was continuously available through the local offices of Peace Energy Cooperative and supplemented through the offices of Aeolis in Sidney, British Columbia.

Aeolis worked to build true partnerships within the community. For example, the signing of a Memorandum of Understanding (MOU) with Peace Energy Cooperative in September 2004 set the path that resulted in PEC becoming a key stakeholder in Bear Mountain Wind with a share in the Limited Partnership. Later, both Aeolis and AltaGas helped ensure that Peace Energy Cooperative remained an important part of the project and received financial benefit from it for the entire 25 year period of operation. Through this process the cooperative, a regional grass-roots initiative, grew, prospered and solidified its influence as a proponent of future renewable energy projects.

Components of the EA

The EA included an analysis of the potential direct and indirect effects of the project on a number of valued ecosystem and social components:

- Atmospheric environment
- Geology and terrain stability and soil
- Hydrology and water quality
- Vegetation
- Wildlife and wildlife habitat, including butterflies, amphibians, bats, breeding birds, raptors, birds on migration, ungulates and fur bearers
- Fish and fish habitat
- Transportation
- Archaeology

• Land and resource use

• Socio-economics and health, including visual impact, shadow flicker and sound

• First Nation and Kelly Lake community's culture and heritage resources.

Pre-construction studies were conducted in 2005 and into the autumn of 2006 and included effects from construction, operation and decommissioning as well as baseline studies of existing data. (Decommissioning of the project is not anticipated as it is expected that after the 25+ year operation period the project will be refurbished and continue to operate). The resulting 400-page document (plus 200 pages of appendices) stands as an in-depth and comprehensive resource detailing the natural and cultural heritage of the Bear Mountain micro-region, and sets a very high standard for public consultation. To obtain environmental certification, AltaGas agreed to both pre- and post-construction surveys and ongoing monitoring programs.

It also set a high standard for low-impact project development. During the detailed engineering phase, the effects of the turbine sites were minimized by locating them where there was a lower potential to damage the terrain. Raptor flight patterns

were taken into account – turbines were set back an additional 50 metres from the ridge to avoid an updraft area that raptors frequent for hunting. Site excavation, adjacent work areas and spoil piles were engineered to minimize erosion, contamination and impact to vegetation. Archeological and First Nations sacred areas were identified and roped off as "no work" zones. Tree clearing was stopped except by special permit during nesting and breeding season, and clearing was kept to a minimum by using existing access roads.

Pre-construction bird, raptor and bat studies identified the types and numbers of species in the area, and the level of bird and bat activity near each turbine site. Monitoring started in the fall of 2007, using mist netting, call playback, visual surveys, radar and acoustic detectors, and will continue for two years after commissioning. Evaluation of results may then lead to more monitoring and/or mitigation measures.

Original plans called for 60 two-megawatt turbines, later changed to 34 three-megawatt turbines. This reduced the total footprint of the project from the original 35 hectares to 25 hectares, less than two per cent of the area covered by the investigative use permit. Areas impacted during construction but not

Above and left page: bird and animal studies were an important part of the EA. Bird counts, migration patterns and nesting locations on the ridge were carefully mapped using many methods including radar and acoustic detectors, call playback, mist netting and visual surveys. Bird, bat and raptor studies will continue for at least two years after commissioning.

required during operation are being restored to their pre-disturbance state using native species. Most of the heavy construction was undertaken in the fall and winter to take advantage of the dry and frozen conditions to reduce rutting and the alteration of surface water flows. It also ensured

that migratory birds were not in the area, several animal species were in hibernation and vegetation was in its dormant stage. During construction, environmental monitors were on-site ensuring that values and concerns identified during the EA were addressed.

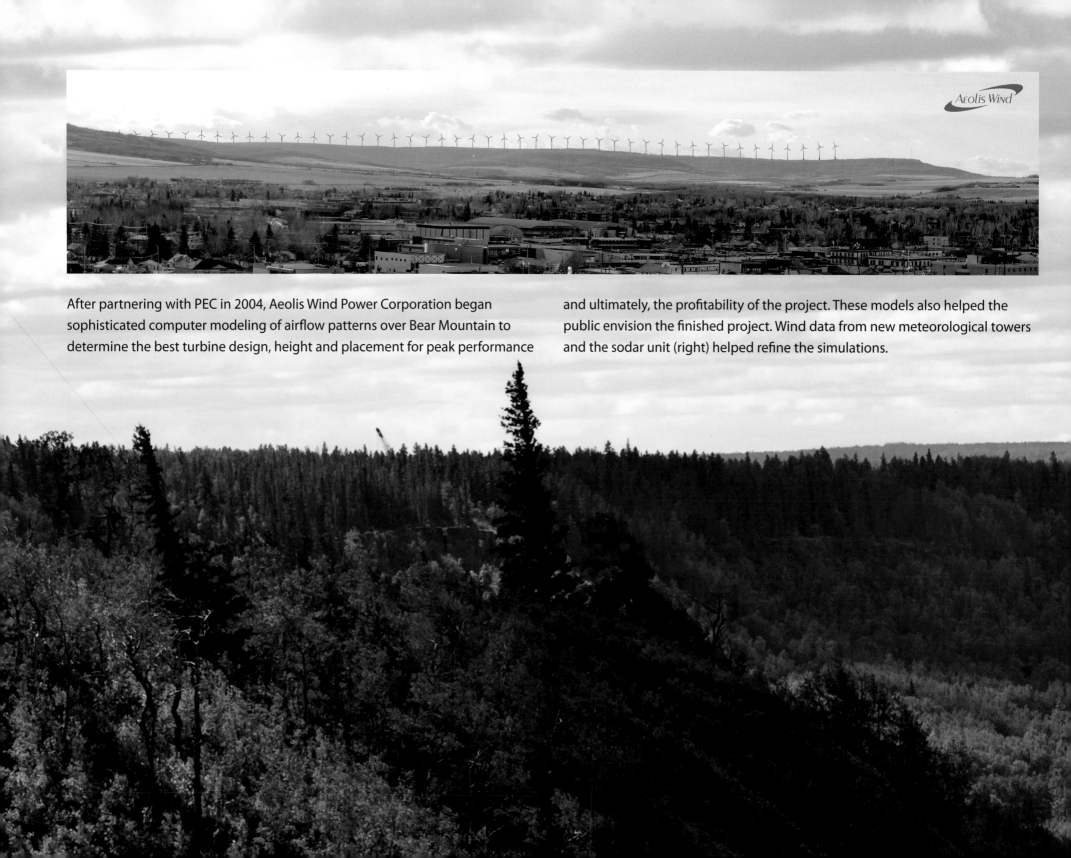

After partnering with PEC in 2004, Aeolis Wind Power Corporation began sophisticated computer modeling of airflow patterns over Bear Mountain to determine the best turbine design, height and placement for peak performance and ultimately, the profitability of the project. These models also helped the public envision the finished project. Wind data from new meteorological towers and the sodar unit (right) helped refine the simulations.

Left: *an early simulated view looking west across Dawson Creek shows many more turbines than were ultimately installed.*

Far right: *because of the critical importance of accurate and reliable wind data, three 60-metre meteorological towers (met towers) were erected at carefully chosen locations on Bear Mountain in 2005 and 2006. The data streams were used to model wind flow over the ridge at different heights.*

Right: *a sodar (sonic detection and ranging) unit was also used to profile the wind resource. Sodar units project acoustic pulses vertically and then listen for the return signal. Wind speed, direction and turbulence can be determined by analyzing the intensity and the Doppler (frequency) shift of the return signal. A profile of the atmosphere at various heights up to several hundred metres can be acquired. Sodar units have many advantages: they are quick and easy to set up and move, they are not physically restricted in height like towers and can be very cost effective. However, at this time sodar data is considered less reliable than met tower data.*

Background image: *the west face of Bear Mountain looking south, before construction began.*

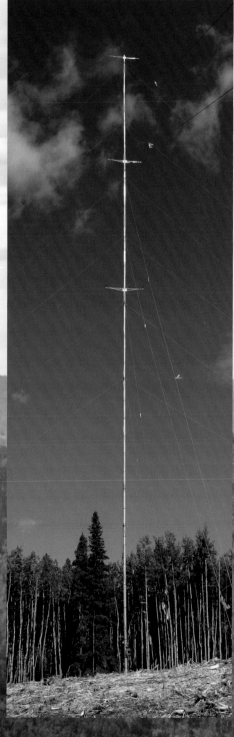

PREPARING the SITE

There were three project managers for the construction of the $200-million Bear Mountain Wind Park: AltaGas Income Trust, ENERCON Canada Inc. and Hatch. AltaGas, as owner of the project, was overall project manager. ENERCON, designer and manufacturer of the wind turbines, was in charge of turbine assembly and commissioning. Hatch managed engineering, procurement and construction of roads, pads, the sub-station, power collection systems and the transmission line.

ENERCON has installed more than 10,000 turbines world wide, but this was the first wind project for AltaGas, and Hatch's fifth. The project was completed on-time and on-budget.

Above: Dan Perrin, Hatch construction manager for the project, leads the kick-off meeting in Dawson Creek, December 2007, covering the overall scope of work and safety requirements. "Our first priority is safety. Our second priority is respecting the environment," said Perrin.

Right: a map of the project showing precise placement of the 34 turbines, sub-station, transmission line and access road.

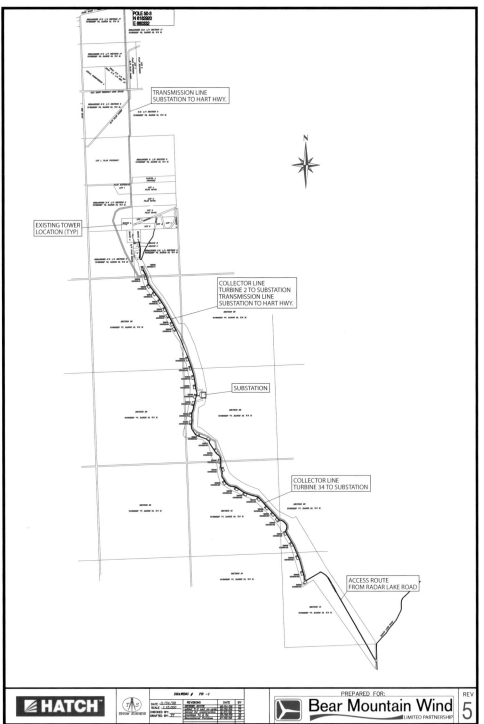

Above: surveyor Geof Tryon of Tryon Land Surveying Limited, Dawson Creek, with his GPS real time kinematic survey unit. Most GPS units are accurate to a few metres, but this one is accurate to a few centimetres.
Previous page 33: operator Bob Patterson (left) and Kevin Moen, one of the owners of Michalchuk Bros. L. Contractors Ltd. in front of an Iron Wolf slasher.

Above top: first ground was broken to begin the construction phase of Bear Mountain Wind Park in mid-December 2007. Dan Irving (centre), construction site superintendent, discusses layout details with surveyors Ern Hegglun (left) and Geof Tryon. At the time of this photo, access was by snowmobile only.
Bottom: tools of the trade – map and GPS.

Above: *continuous environmental monitoring was an important part of construction operations on Bear Mountain. Doug Russell, Green Tree Resource Contracting Ltd. of Dawson Creek, checked "no work" zones around sensitive areas and archeological sites to be sure they were well-marked and respected. He filed daily environmental reports.*

Right: Marc Boucher, project manager for AltaGas, directed operations as the first pioneer road was roughed in along the ridge. All equipment carefully followed an accurately surveyed and flagged route to meet EA minimum impact requirements.

Left page: *after logging, residual organic material from roadsides, pads and rights-of-way was mulched and used as soil cover to help accelerate regrowth. Bob Patterson, heavy equipment operator for Michalchuk Bros. L. Contractors Ltd. of Grande Prairie, Alberta takes a break for this photo.*

Above: *aspen trees removed for roads and pads were carefully cut and prepared to specific standards for acceptance at the LP Oriented Strand Board (OSB) plant near Dawson Creek. Logging was not allowed during the spring bird and wildlife breeding season.*

Left page, top left to bottom right: *Brandi Bronowski, Mountainview Safety Services signing in Marc Boucher, AltaGas project manager; GeoNorth Engineering drilling and soil testing crew; Dan Irving, construction site superintendent; Brandi Belton, truck driver, Brocor Construction; Calvin Napoleon of Flaming Arrow Contracting; and Mary Anderson, compactor operator, Brocor Construction.*

Above: *in places, road building and pad clearing required moving a lot of overburden. Mulched debris and topsoil can be seen piled out of the way, ready to be spread over the finished contours to help speed vegetation regrowth. Overall, some 400 people worked on Bear Mountain Wind Park, 300 on-site. About 100 were locally hired and an estimated $10 to $20 million dollars flowed into the local economy.*

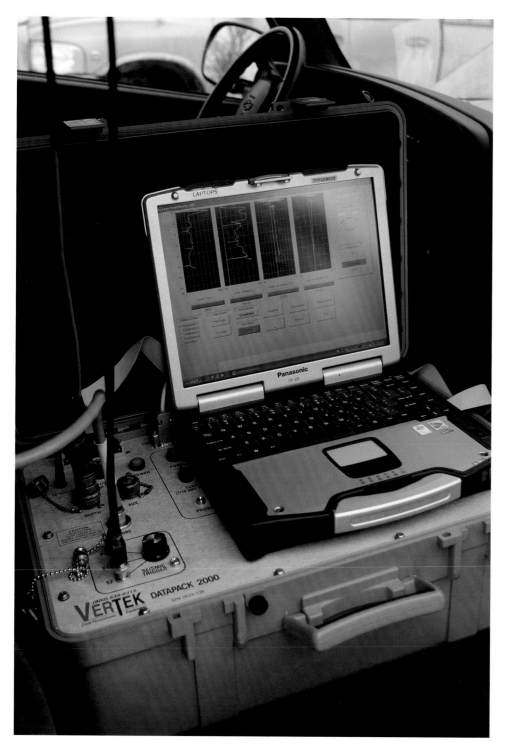

This page: *each tower location underwent site specific drilling to analyse sub-surface composition and determine the most appropriate foundation design parameters for that location.*

Left page: *Brocor Construction of Dawson Creek was the principal site preparation contractor.*

From top left to bottom far right: *this sequence of images shows progress from mid-December 2007 just after ground breaking, to site preparation completion in late summer of 2008. In the image at bottom right on the next page, carefully contoured*

banks have been stabilized with erosion control blankets. Even with the extensive site preparation needed for Bear Mountain Wind Park, the time from breaking ground to final completion and commissioning was less than two years. After commissioning,

most of the project footprint will be reseeded with native species and a forage seed reclamation mix that will expand cattle grazing capacity, a traditional use of this area that is highly compatible with wind power. Because the actual footprint

of the project is so small (less than two per cent of the original investigative use permit area), recreation, grazing, forestry and most other existing activities on Bear Mountain should be unaffected or somewhat enhanced.

Building an ENERCON E-82 wind turbine foundation. Step 1: after soil testing and surveying, remove overlying organic soil or glacial sediment and excavate down to compact till or bedrock to a diameter of about 18 m. *Step 2:* install conduits for underground power and communications lines. *Step 3:* pour a thin cement working pad and lift the foundation tower section (green) into place. *Step 4:* add 35 tonnes of foundation reinforcing steel and copper grounding wire. *Step 5:* complete the

complex reinforcing steel structure. **Step 6**: attach two sets of circular steel forms and pour 300 tonnes of concrete, about 50 truckloads, into each foundation. **Step 7**: let the foundation cure and remove the steel forms. **Step 8**: backfill. The

mass of compacted backfill is part of the calculated foundation structure, helping to reduce foundation size, mass and cost. Jay Walker, construction site superintendent, gives a sense of scale.

Left and facing page: a complex network of high-strength steel reinforcing bar is carefully woven throughout the foundation structure. The rebar comes from the turbine manufacturer pre-formed to fit an exact pattern, like a giant jigsaw puzzle. Specially trained teams of expert foundation rebar assemblers travel from site to site around the world, quickly and efficiently assembling this critical part of the turbine foundation.

Right top: this foundation is ready to be filled with concrete. The intricate network of rebar and the circular steel forms, top and bottom, are in place.

Right bottom: the tower collar (seen here in the background wrapped in green plastic to protect it from concrete splatter) will have a 78-metre, 214-tonne steel tower attached directly to it, so it has to be levelled and positioned to less than a millimetre of accuracy.

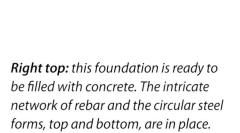

Above: *completion of individual foundations was occurring on a weekly basis when officials arrived for a site tour and photo opportunity in August 2008. Left to right: Dr. Chris Garwah, PEC board treasurer; Valerie Gilson, PEC executive director; Calvin Kruk, mayor of Dawson Creek; Jim Bracken, AltaGas senior vice president, major projects; Steve Rison, PEC president.*

Above left: *concrete is poured into the foundation tower section, firmly embedding it into the rebar foundation structure.*
Above right: *samples of each batch of concrete are tested before and during the pour to ensure consistent high quality.*

Right page: *foundation concrete must be poured continuously, without interruption, to ensure highest possible strength. The approximately 50 truckloads of cement needed for each foundation arrive "just-in-time" to pour their contents into the pumper truck (white with red design). The pumper truck is remotely controlled by an operator (yellow hard hat, second from left) who watches the pour closely and regulates flow.*

Aerial photos: by September 2008, the leaves on Bear Mountain had just begun to turn. At left, notched clearings for 34 towers can be seen along the nine-kilometre access road. Above and at far right, completed and cured foundations are being covered with earth. Through the fall and winter, roads and tower clearings were completed and graveled for the arrival of turbine and tower components in the spring. Power lines and the substation were also finished through the winter months.

Above: as foundation work was nearing completion, AltaGas officials visited the site in mid-September. Left to right: Marc Boucher, project manager; David Cornhill, chairman and CEO; Jim Bracken, senior vice president, major projects.

ASSEMBLING the TURBINES

Previous page 55: *the first shipment of 10 turbines (blades and major components) left Emden, Germany in April 2009 aboard the MV Maxima. The photo shows her moving through the Welland Canal, with the 41 metre blades stowed on deck, on her way up the Great Lakes to a May 7 arrival in Thunder Bay. The towers, manufactured in Saskatchewan, were shipped separately by truck.*

Above and right: *at Thunder Bay the blades were off-loaded and shipped across Canada by rail to arrive in Dawson Creek during a late-May snow storm. The other components were sent from Thunder Bay by truck.*

There was a sense of excitement, of "making history," throughout the Bear Mountain Wind project, but especially when the first large components began to arrive in Dawson Creek. In this region of farm machinery and oil rigs, the first shipment of blades, so long, sleek and otherworldly, were the talk of the town.

Special steel cradles that bound the blades in pairs made it possible to handle them many times without damage during transport: stacked on ship decks (page 55), secured to flat bed rail cars, lifted by crane from rail to truck (right page), transported by truck to the holding yard near Dawson Creek (pages 60 - 61), lifted again onto trucks for the short ride up to Bear Mountain, and finally lifted from the delivery truck to the ground at the construction site, all without leaving their cradles.

Above: the Canadian-designed and made blade-hauling trailers shown here are powered independently and steered from the cab. **Left:** a hundred blades in temporary storage at the Brocor Construction yard near Dawson Creek.

This page: *power line and substation work were in full swing through the early months of 2009, preparing for "first power" later in the year.*

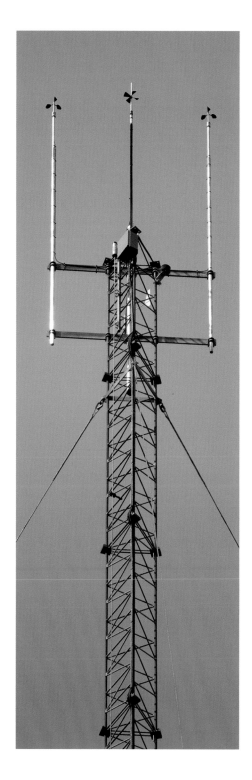

This page: *in May 2009, a new 80-metre meteorological tower was installed on-site. This permanent tower, with wind instruments at turbine hub height, will be used by facility controllers to fine tune the less-accurate wind data collected from individual turbine sensors.*

Above: *climber Terry Bouthillier of Waybest Tower Inc., Medicine Hat, Alberta, has tremendous respect for, but very little fear of heights.*

Above: *May in the Peace River region – fragrant aspen buds were turning into bright green leaves as the first major turbine components arrived on-site. The first tower section arrived from Saskatoon, Saskatchewan on a unique rig designed specifically for this purpose.* **Right page top left:** *the two sections of the truck rig are seen rejoined for the return trip after the tower section has been removed.*

Right page top right: *the E-module (silver) has been placed inside the foundation tower section (green ring) ready for the next tower section to be placed over it.*
Right page bottom row: *crews prepare to lift the first section of the first tower. In these first lift photos, there are more personnel than normal because many are being trained for the six months of turbine assembly ahead.*

Cranes

Building modern wind parks would not be possible without technologically sophisticated crane equipment and skilled operators. Two of the world's biggest cranes, a 440-tonne "narrow track" crawler crane and a 440-tonne standard crawler crane, plus six smaller cranes (100 to 300 tonnes) were used on Bear Mountain. The narrow track crane is the only one of its kind in Canada. It allowed the roads to be just 5 m wide rather than the usual 10 m, greatly reducing the environmental footprint. The narrow roads and small working pads required by the EA limited on-site storage, so that major components were delivered "just-in-time" then lifted into place. This also meant that the 440-tonne standard crawler crane had to be disassembled to move it from pad to pad. Eagle West Wind Energy, based in Calgary, Alberta and Abbotsford, B.C., was the primary crane contractor for Bear Mountain Wind Park, supplying the eight cranes and more than 50 installers needed to erect the turbines.

These pages: the first tower section is lifted to enclose the silver E-module, an electronics module that contains most of the turbine control and monitoring systems. The wind turbine generator feeds AC power at 400 V to transformers in the E-module that convert it to a medium voltage of 34.5 kV for distribution to the on-site collector substation.

This bottom tower section, painted the distinctive ENERCON green, is 12.35 metres long, 4.5 cm thick at the base and weighs 61 tonnes.

Photo sequence at right: *tower section two (18.2 metres long and weighing 55 tonnes) is placed by two cranes working together.*

Bottom sequence: *the third tower section (21.8 metres, 48 tonnes) is placed by the 440-tonne narrow track crawler crane assisted by a smaller crane.*

Below: *a close up of the 440-tonne standard crawler crane.*

This spread left to right: *the massive 440-tonne crawler crane dwarfs crew members on the ground; workers and equipment are lifted into place when it becomes impractical to climb the inside tower ladder; a distant view of the crane moving a 24-metre, 35-tonne top (fourth) tower section into place; placing the third tower section, 21.8 metres long and weighing 48 tonnes.*

Above: *workers guide the crane operator by radio, then slip temporary pins into the bolt holes around the tower flanges once they are aligned. A ring of bolts is then installed and tightened to join the two sections permanently. There is no* gasket at the joint – the two polished flanges fit so perfectly, and are so tightly bolted, they form a weather-proof seal.

High crane lifts have to be done during times of calm air, a challenge in a location chosen for its strong winds. Large cranes like these are fitted with sensors that automatically shut down the crane should unsafe winds suddenly occur.

Previous page 70: *workers are at the ready as the red crane lifts the next tower section towards them.* **Page 71:** *a wide-angle view distorts perspective but adds drama to this crane and tower silhouette.*

Left page: *once the nacelle is lifted 78 metres to the top of the tower, the crew is lifted to the nacelle. The nacelle forms a sheltered work space for future maintenance, and houses cooling and heating equipment, wind sensors and the motors and gears needed to rotate the entire nacelle/generator/blade assembly into the wind.*
This page: *a 53-tonne three-megawatt generator is mated to the nacelle. It has two* main parts: the inside "rotor" that is driven in a slow rotation by the blades, and the outside stationary "stator." There are no gears or other fast-rotating parts in the ENERCON power system, so gear oil, noise, mechanical wear and energy loss are reduced. **Lower right:** *ENERCON field service engineer Viktor Kisser waves from on high. Viktor and his family came from Germany and now live near Dawson Creek.*

This page: *the hub is prepared to be lifted into place in front of the generator. Once in place, the blades will be attached. The 24.5-tonne hub assembly contains motors and gears that can rotate each blade in or out of the wind to control rotation rate and power output.*

This page: *looking like visitors from another planet, assembled turbines await their blades and calm days for lifting them. The ENERCON modular turbine design has just three components: the hub, the generator and the nacelle. This modular concept, in conjunction with a gearless design, reduces maintenance and down time.*

Above and left: the first blade was lifted into place July 14, 2009. Each blade is 41 metres long, weighs eight tonnes and is made of epoxy resin with integrated aluminum in the tip and along the leading and trailing edges for lightning protection. At many wind facilities, blades are attached to the hub on the ground and lifted as a unit, but Bear Mountain's restricted working area meant blades had to be lifted one at a time once the hub was in place at the top of the tower. Very calm conditions are needed for blade lifts.

Blades are "variable pitch" meaning they automatically respond to wind speed variations by rotating in and out of the wind, independently yet synchronously, to maximize power and minimize stress on the system. To shut down the turbine, the blades are rotated out of the wind. Within a few seconds the rotor goes into a very slow rotation known as "idle mode" which places lower stress on the drive train than does a locked rotor. The rotor lock is used as a final safety measure during maintenance procedures.

Far left: this end view of an ENERCON blade shows the blade profile extending right down to its base where it flares to capture inner air flow around the streamlined nacelle. The distinctive bent tip extends the power capture field to the very end of the blade, minimizes wind sound and improves rotor blade stability.

These pages: cranes continued lifting blades into place even as the first completed turbine at Bear Mountain Wind Park was activated on July 23, 2009 (lower right) and wind power began feeding into the British Columbia electrical grid for the first time. Witnessing the moment of first power were, left to right, Lizie Dunling-Smith, AltaGas' project engineer; Jay Walker, construction site superintendent; and Marc Boucher, project manager for AltaGas. This turbine was officially commissioned (tested and declared fully operational) about one month later.

One of Canada's Largest and Fastest Growing Energy Infrastructure Organizations

August 6, 2009: the official launch of British Columbia's first wind energy facility, Bear Mountain Wind Park, attracted significant media attention.

It was a first for everyone: BC Hydro, AltaGas, Peace Energy Cooperative and the province. The City of Dawson Creek was supportive from the beginning and saw it as a perfect fit with its "Sustainable Dawson Creek" initiatives that were attracting national recognition.

Far left: visiting dignitaries pose for a commemorative shot. All levels of government were involved in the project.

Speakers, top left to bottom right: Hon. Richard Neufeld, senator for B.C.; Jay Hill, MP for Prince George-Peace River; Barry Penner, B.C. minister of the environment; Blair Lekstrom, MLA Peace River South and B.C. minister of energy, mines and petroleum resources; Mike Bernier, mayor of Dawson Creek; Steve Rison, president of Peace Energy Cooperative.

Right: David Cornhill, AltaGas chairman and CEO, handed out "wind power cake" during the celebration.

This page: the on-site substation was quickly ramping up power to the grid as each of the 34 turbines came on line, peaking at 102 megawatts at full commissioning in October 2009. Individual wind turbines supply alternating current at 400 V to transformers located in the E-module within the base of each tower. These transformers convert the output power to the medium distribution voltage of 34.5 kV which is stepped up to the transmission voltage of 138 kV in the substation. Power is then fed to the grid via a 4.3-kilometre overhead transmission line heading north to the regional main 138 kV power line running between Dawson Creek and Chetwynd, parallel to the Hart Highway.

Left page: in these aerials from September 2009, several cranes work simultaneously along the line, erecting the last towers and lifting the last blades.

It was harvest time in the Peace Country, and the region's first wind facility was quickly reaching full operating capacity. As each turbine's performance was tested and confirmed, ownership passed from ENERCON to AltaGas. Day-by-day, more and more truly green electricity flowed into the provincial grid for the first time ever, enough to power the entire South Peace, more than 35,000 households. As blades began to turn, local pride began to soar. Many had literally made history by helping build the project and hundreds more had invested time and money through AltaGas and Peace Energy Cooperative to see British Columbia's first wind park up and running.

The FUTURE

Careers in wind

The fastest growing source of electrical energy in the world is also growing jobs at a phenomenal rate. Demand for trained personnel is rising dramatically right alongside the increase in wind utilities worldwide.

"There is tremendous opportunity, especially if you want to travel," explains Howard Mayer, dean of business, industry and contract training at Northern Lights College, the regional college based in Dawson Creek. Northern Lights was the first in British Columbia to offer a one-year Wind Turbine Maintenance Technician program and the BZEE certification – the European training standard for wind turbine technicians. Graduates can move directly to a job in the field, then progress to higher levels of certification within the industry, often assisted by specialized training provided by turbine manufacturers. Career paths are varied and include mechanical, electrical, supervisory and instructional opportunities.

"Small wind" is also expanding as individual homeowners, ranchers, farmers, businesses, schools and municipalities begin installing independent energy systems that include wind generation.

Top: *Northern Lights College student Luke Potosky is learning high angle rescue as part of the Wind Turbine Maintenance Technician program.*
Left and above: *Joel Faulkner (left) and Syd Paschen are one of two, two-person electrical service teams who, along with a two-person mechanical crew and operations supervisor, keep Bear Mountain Wind Park in top operating condition.*

Above left: AltaGas' operations supervisor Don Hallam checks the real-time status of all 34 turbines on his SCADA (Supervisory Control and Data Acquisition) system at the on-site substation. The SCADA system provides secure internet access and continuous monitoring from the ENERCON Nova Scotia service centre, ENERCON in Aurich, Germany, AltaGas headquarters in Calgary and Don Hallam's home.

Above: "I love coming to work every day," says electrical service technician Syd Paschen. "It's clean, it's green, it's really interesting. There is unlimited opportunity all over the world." *Above left:* Syd adjusts a wind sensor atop the nacelle. *Above right:* strapped to the nacelle 80 metres above ground, Syd takes a moment to enjoy the feel of wind on his face, a reminder of the great power harnessed by these machines.

The future of wind power

As this book goes to press, data from around the world strongly indicates that dramatic climate change is upon us, well ahead of predictions. Energy, how we produce it and how we use it, underpins all of our activities, and by far contributes the most to green house gas (GHG) production and environmental pollution. GHGs are the single greatest factor in the climate change equation. Moving to a clean energy world is the most pressing challenge we face today.

Wind is the most environmentally friendly source of electric power. Wind is the fuel, so wind energy facilities do not generate pollutants or hazardous waste. Fossil fuels such as coal, oil or gas are minimally required for the manufacture, transportation and installation of turbine components, greatly reducing environmental damage through resource extraction and transportation.

Nuclear power results in up to 25 times more carbon emissions than wind, when reactor construction and uranium refining and transport are considered. Carbon capture and sequestration can reduce CO2 emissions from coal-fired power plants but will *increase* air pollutants and the other problems of coal mining, transport and processing, because significantly more coal must be burned to power the capture and storage steps.

A wind facility generates only one per cent of the CO2 emissions produced by coal fired electricity, and only two per cent, per unit of electricity, generated by a natural gas fired plant. These numbers are based on the *total* fuel cycle, which includes the manufacture of equipment, construction of the facility and production of electricity. In other words, using wind instead of coal reduces CO2 emissions by 99 per cent, and using wind instead of gas reduces CO2 emissions by 98 per cent.

The net annual GHG emission reduction resulting from the operation of Bear Mountain Wind Park will be about 280,000 tonnes of CO2 per year. That is a GHG CO2 reduction of more than seven million tonnes over its 25-year operating life.

Even though wind energy has a high level of public acceptance in British Columbia (a 2006 study by BC Hydro revealed a 98-per-cent approval rating), by global standards, wind energy development in the province has been slow to get underway. With an estimated 10,000 MW of potential wind power available in the British Columbia Peace region alone, where the province's largest hydro electric dam, the W.A.C. Bennett Dam is located, the province is well positioned to take advantage of the synergistic mix of wind and hydro.

As the cost of cleaning up conventional energy sources goes up, the cost of renewables continues to fall. The cost of wind energy production has decreased by about three per cent per year for the past 25 years due to technological improvements in wind turbine design. Wind power is already cost-competitive with all other energy sources even without factoring in environmental impacts. Soon it will be the least expensive of all energy options.

Europe is the leader in wind power. Denmark produces over 25 per cent of its electricity from wind. Interconnection between Denmark and Norway, which has hydroelectric generation, provides a firm and reliable supply by combining wind energy with the storage capacity of hydro. Denmark intends to obtain over 50 per cent of its electricity from wind, and Britain recently approved a program to develop 7,200 megawatts of new wind energy. To compare, the total generating capacity of British Columbia's utility, BC Hydro, is estimated to be 10,000 megawatts from all sources.

Despite a widespread economic recession, new wind power capacity grew more than 31 per cent in 2009, the highest rate in the last eight years, to a total of 158,505 megawatts worldwide. China surpassed the United States to become the world's largest wind turbine market. Wind power produced 340 trillion kilowatt-hours, or two per cent of total global electricity consumption, in 2009.

The most recent studies indicate it is possible to more than meet *all* our global energy needs with a combination of renewable resources: wind, solar electric, solar thermal, hydro and geothermal. The cost of generating and transmitting that power would be less than the projected cost of generating it with fossil fuel and nuclear power.

How much would be required? A 2009 study in Scientific American, *A path to sustainable energy by 2030*, by Mark Z. Jacobson and Mark A. Delucchi, proposes a plan to eliminate the need for all fossil fuels by 2030 using a mix of 90,000 solar plants, numerous geothermal, tidal and rooftop photo voltaic installations, and 3.8 million five-megawatt wind turbines worldwide. (Wind supplies 51 per cent of world demand in this plan because it is the fastest to scale up and deploy.) This seems like a lot of wind turbines, and it is, but lets remember that around the globe, 73 million cars and light trucks are manufactured *every year*.

A rapid transition to a clean energy world is not only possible, but critically necessary. For our grandchildren, yes, but also for ourselves. We have unwittingly created a new planetary environment, one in which climate is changing in unpredictable, far-reaching and fundamental ways. Not in some distant future, but now, all around us, affecting every living thing on the planet. And the changes have only just begun.

We have the means – the technology and the skills – to transform our energy economy and by so doing, our world. It is clear that wind power is a crucial part of the solution and will be a significant part of the world our children inherit. For the people of the South Peace region of British Columbia, Canada, Bear Mountain Wind Park is a symbol of hope and inspiration. For the world, it is a perfect example of the POWER SHIFT, and a blueprint of how to do it well, do it right, and do it now.

Page 89: the nature trail along Bear Mountain ridge offers superb views to the distant mountains, and an up-close view of a state-of-the-art wind park. These pages: spring of 2010 brought the first school tours to Bear Mountain Wind Park, and lots of "ooh's and aah's" too. About 50 students and teachers from Dawson Creek's Crescent Park and Tremblay elementary schools, grades 3 to 6, got a close-up look at wind energy, plus presentations from Aeolis and Peace Energy Cooperative.

Dawson Creek in November 2009, the first city in British Columbia to be flooded with clean, green electrons generated by the wind. Bear Mountain Wind Park, seen silhouetted on the distant ridge, produces enough power for 35,000 homes or more than three complete cities the size of Dawson Creek.

POWER SHIFT was written, designed and laid out using at least
70 per cent photovoltaic power.

Peace photoGraphics is dedicated to publishing books and other media
that inspire an appreciation of the natural world and our place within it,
especially in the Peace River bioregion of northwest Canada.
We make every effort to be socially and environmentally responsible,
recognizing that a healthy planet and a healthy human society
go hand in hand, and that we all have an important role to play
in making a better world.

Our complete line of books, cards, photographs and DVDs can be found at
www.peacephotoGraphics.com

Phone: 250-782-6068 or toll free 866-373-8488
Fax: 250-782-9133
Email: info@peacephotoGraphics.com

Peace photoGraphics Inc.
1204 - 103rd Avenue
Dawson Creek, British Columbia
Canada V1G 2G9